I0813735

ALL ABOUT

BABY FLAMINGOS

by Martha E. H. Rustad

PEBBLE
a capstone imprint

Pebble Emerge is published by Pebble, an imprint of Capstone.
1710 Roe Crest Drive
North Mankato, Minnesota 56003
www.capstonepub.com

Library of Congress Cataloging-in-Publication Data
Names: Rustad, Martha E. H. (Martha Elizabeth Hillman), 1975- author.
Title: All about baby flamingos / by Martha E.H. Rustad.
Description: North Mankato, Minnesota : Pebble, [2022] | Series: Oh baby! | Includes bibliographical references and index. | Audience: Ages 5-8 | Audience: Grades K-1 | Summary: "There's a new baby joining the flock. It's a flamingo chick! Learn all about baby flamingos, including what they eat, what they weigh, how they're raised, and how big they grow"—Provided by publisher.
Identifiers: LCCN 2021002740 | ISBN 9781663907875 (hardcover) | ISBN 9781663907844 (pdf) | ISBN 9781663907868 (kindle edition)
Subjects: LCSH: Flamingos—Infancy—Juvenile literature.
Classification: LCC QL696.C56 R87 2022 | DDC 598.3/51392—dc23
LC record available at https://lccn.loc.gov/2021002740

Image Credits
Getty Images: picture alliance, 15; Shutterstock: Anand RJ, 18, David Ryo, 9, GUDKOV ANDREY, 13, Henner Damke, back cover, Jeff Kingma, 5, Ondrej Chvatal, 11, 17, PrimaStockPhoto, 21, Stas Malyarevsky, 20, Stephaniellen, cover, 6

Editorial Credits
Editor: Alison Deering; Designer: Jennifer Bergstrom; Media Researcher: Tracy Cummins; Production Specialist: Tori Abraham

All internet sites appearing in back matter were available and accurate when this book was sent to press.

Printed and bound in China. 5485

Table of Contents

Words in **bold** are in the glossary.

A FLAMINGO CHICK

Cheep! A new baby is here. It is a baby flamingo. Baby flamingos are called **chicks**.

The chick hatched from an egg. Its mother and father took turns sitting on the egg. They had to sit on it for about a month.

The chick lives in a nest on the ground. Its mother and father built the nest near water. They used mud.

The chick is not strong at first. It slowly learns to walk. It falls over a lot while learning.

The mother and father take turns feeding their baby. Their bodies make a red liquid. This is called milk. It drips from their beaks. It goes into the chick's open mouth.

GROWING UP

A flamingo is not born pink. Its feathers change as it grows. A baby is fluffy and white. By 1 month old, it has soft gray feathers.

The chick looks different at 3 months old. Black feathers start to grow on its wings. It starts learning to fly.

A flamingo chick has a straight **bill**. But it changes shape. It becomes curved. It is time for the chick to feed itself.

Flamingos eat tiny animals that live in water. The birds walk in shallow water on their long legs. A flamingo's legs can be 30 to 50 inches (76 to 127 centimeters) long.

Flamingos reach their long necks to the water. Their curved bills scoop up water. **Bristles** on their beaks help them **filter** food.

FLAMINGO GROUPS

Flamingos live in large groups. These are called **colonies**. Some have hundreds of birds. Some may have thousands.

Flamingos travel together. They fly long distances. They search for food and water. They all build nests in the same place. They raise their babies together.

Flamingos honk, grunt, and growl. Each family has its own set of sounds. The mother and father teach them to their chick while it is still inside the egg.

A baby flamingo stays in its nest for about a week. Then it leaves. It joins a small group of chicks. Each baby goes back to its parents to eat. It listens for a call from its parents. Time for dinner!

After a few weeks, the adults gather the chicks into a large group. This group is called a **crèche**.

ALL GROWN UP

Young flamingos are fully grown around age 3. But they are often still white or gray. It can take up to three years for their feathers to turn pink.

Their color comes from two places. Their food makes their feathers pink. It also comes from the red milk they ate as chicks.

A flamingo finds its **mate** around age 6. A group of male flamingos move together in a dance. They show off their feathers to females.

A male and female flamingo pair up for life. Flamingos live for 40 to 60 years.

STAND LIKE A FLAMINGO

A flamingo often stands on one leg. Scientists think they do this to stay warm. At least they can keep one foot out of the cold water! How long can you balance on one foot?

What You Need

- your shoes
- a wall
- a timer
- a friend or adult to help

What You Do

1. Put on your shoes. Stand next to a wall. Practice standing on one foot. If you lose your balance, you can touch the wall. Ask your helper to time how long you can stand without touching a wall.
2. Try standing on the other foot. Is it easier or harder? Ask your helper to time you.
3. Now take off your shoes. Try standing first on one foot and then the other.
4. Now try closing your eyes and standing on each foot alone. Is it easier or harder?

Glossary

bill (BIL)—the hard front part of the mouth of birds; also called a beak

bristle (BRIS-uhl)—a short, stiff hair

chick (CHIK)—a young bird

colony (KAH-luh-nee)—a large group of animals that lives together

crèche (kreysh)—a small group of flamingo chicks

filter (FIL-tuhr)—to move through or into something in small amounts or in a gradual way

mate (MATE)—one of a pair

Read More

Dickmann, Nancy. *Why Are Flamingos Pink?* Mankato, MN: Capstone, 2021.

Klepeis, Alicia Z. *Flamingos*. Minneapolis: Jump! Inc., 2021.

Riggs, Kate. *Flamingos*. Mankato, MN: The Creative Company, 2015.

Internet Sites

National Geographic Kids: Flamingo
kids.nationalgeographic.com/animals/birds/flamingo/

How Do Baby Flamingos Become Pink?
thekidshouldseethis.com/post/how-do-baby-flamingos-become-pink

San Diego Zoo Kids: Flamingos
video.link/w/8fqmb

Index